INSIDE CLIMATE CHANGE

INSIDE CLIMATE CHANGE

The Book Of Facts, Poems, Riddles and Rhymes

Professor What-If

INSIDE CLIMATE CHANGE
The Book Of Facts, Poems, Riddles and Rhymes

iUniverse books may be ordered through booksellers or by contacting:

iUniverse
1663 Liberty Drive
Bloomington, IN 47403
www.iuniverse.com
1-800-Authors (1-800-288-4677)

Cover illustration by Barbara Ann.

ISBN: 978-1-4917-9052-6 (sc)
ISBN: 978-1-4917-9053-3 (e)

Library of Congress Control Number: 2016904034

Print information available on the last page.

iUniverse rev. date: 08/19/2016

Reform is a thin line between costly change and negative resistance.

—Professor What-If

To my grandmother Jessie, who taught me to respect this land; my wife, Barbara Ann, who taught me to always look for the good in people; and the victims of horrific storms and their families.

Contents

Section 5: Transformation

Section 6: Living Underground

Section 7: Professor's Insights

Preface

It was September 2008, and I hadn't realized at the time that a process was occurring inside of me … one that I had mistakenly confused as guilt. I wore this burden as a soldier wears his backpack, with one exception: I could not remove it. It was like a built-in panic alarm always signaling that something was wrong with someone or that something had been left undone. But who? Who was it? Who was in trouble, and what was I supposed to do? At the time, I couldn't distinguish whether this was a warning sign or self-pity. My wife had died, leaving me alone with the responsibility of protecting our children. Although it had been two years since her death, it sometimes seemed like yesterday.

Months later, I woke from what I thought was a sound sleep with symptoms of paranoia. In my sleep, I had been gathering pieces, fragments of this process. Like an artist's early penciled sketch, I gathered small pieces around a larger canvas … black-and-white shapes with just a hint of coloring. This continued for some time as the canvas became occupied with images and color. Color so pure, so deep, unlike anything I had ever seen. It was revealing what I must do. The next morning, I had an awareness of how it should all come together, how to balance our planet Earth, as if I were learning from Mother Nature herself in another dimension.

I know this period of enlightenment is hard to imagine, and yes, I thought I was going mad. I felt foolish thinking about it and would second-guess myself. What is this? Why me? What did I do? Who is in trouble? What is going on? I found myself repeating the same list of questions over and over. At first, the lessons came to me like magnetic fragments gathering in my mind four times a week. They didn't make any sense. I could not formulate the fragments into anything cohesive, somewhat like a stranger talking to me in a foreign language. In the first year, I experienced this process as few as once and as many as four times a week. These episodes varied in intensity.

For the first year, I kept this all to myself as I tried to figure out what was going on. Well into the second year, I occasionally allowed fragments to drift out, sharing them first with my wife and later with close friends. I began to point out things that were about to happen before they did. I would describe these lessons as "I don't hear voices. I just wake up and know!"

Soon I observed a correlation between the levels of intensity of these lessons and the depth of my studies regarding the Gulf oil spill. By years three and four, the connection with super storms, earthquakes, and oil productions caused my lessons to dramatically multiply. By this time, I was spending a great amount of time researching world weather patterns over the past thirty years and world oil production back to 1859, the time of the first oil well.

On Valentine's Day 2014, after hearing a news report that an agreement had been reached between Russia and a major oil company to open Antarctica for oil exploration, I felt sick inside. I knew this was wrong! The greed had to stop. The arguing over climate change had to stop before we totally destroyed our planet Earth. I had to do something; I had to inform the world of what I had learned! But how? Who was I? A person with bad dreams and nightmares? My self-doubt lasted less than ten minutes, and suddenly I knew I had to write this book and let the world judge.

It was on my third attempt to write this book when "she" told me to start over. I was on my fourth revision of the book when I woke up and just knew. I could not write this as a traditional book; it would sound like too much doom and gloom, an angry rant of blame. "She?" you ask. "Who is she?" I like to think of her as Mother, Mother Nature, but I don't really understand why. *I just wake up and know!*

Introduction

Global warming elicits a wide range of emotions, including anger, fear, confusion, and disbelief. Professor What-If shares his insight with an extraordinary blend of informational facts, poems, riddles, and rhymes, revealing the core issues of climate change.

His vision will generate awareness of how serious this problem really is, a problem that all of us, including our children and grandchildren, are facing.

Caution! This is not meant for the emotional kind! This self-proclaimed professor takes you on a poetic journey into what life may become if we continue our ways. It is a guide to open your mind, to probe the question "What if?"

Informational Guide

Professor What-If has intentionally provided three different types of resources to help you navigate *Inside Climate Change*. The following appear periodically throughout the book:

Splash
Researched facts (with sources) to explain the subject matter, along with context to guide you along.

Clash
Commentary and opinions by Professor What-If.

Flash
Professor What-If's insight into what it all means.

A narrative by the professor in section 7 reveals facts about each poem.

1
BALANCE

The Book of Facts, Poems, Riddles, and Rhymes

A look into the window of time
prods your soul,
intensifies your mind.
If not cautious,
you could get left behind!
Journey into this book of time
not meant for the skeptical kind!
Explores climate change
from another side.
Just a glimpse,
but more than enough
from inside out.
Of what is now
and heading our way.
Simple expression of years of greed.
Lack of understanding,
harmony of what she needs
to keep us steady,
balanced, upright.
We share the blame.
We paid no mind.
We had no time!
Now it's here upon us all,
the brink of change.
Our climate is changing!
What's to come?
What's our fate?
Or is it too late?
Trapped in this thirst called greed.
Nowhere to turn.
Nowhere to hide.

You can't run from the window of time!
This is the book of facts,
poems, riddles,
and rhymes.

The Need for Greed

This feeling of greed.
Where did it come from?
Why such a need?
Generations of junkies fixed on greed!
Breeding this addiction.
Expanding our need,
nourishing our desire,
our consumption,
our greed.
How do we tame
this overwhelming need?
Primed to believe our greed is just.
There's plenty,
plenty for all of us.
But at what price?
Much more than pump.
Watching our sky turn to rust.
Destroying our atmosphere
as we debate, why?
Watching our jet stream
dance in the sky.
Our weather is changing.
Just look around.
Much more than
"Greenies" creating noise,
much more than violent storms.
Mother Earth is warning,
"Control your greed!"
Black gold, crude oil.
The ballast of earth!

Mother Nature

Hear that?
What?
That!
What's that?
She's crying.
Who is crying?
Mother Nature, they say.
She's trying to find her way.
Spinning and spinning,
wobbling, wobbling up and down.
Hot into cold, mingling days.
El Niño drawing closer, expanding in size.
Such violent weather is heading our way,
season into season, changing around.
She's out of control,
waffling, waffling.
Day becomes night.
She's failing this fight!

Losing Ballast

It continues day after day,
night into day,
a continuous flow of crude oil.
"Black gold," I heard someone say.
"Her blood," another man whispered.
Around the world in every way!
Large syringes penetrating her layers.
Raping her night and day!
Extracting her gold, her blood,
ballast of earth.

As we build manufacturing monsters,
consumers of greed,
for things they say we need!
Spewing their waste,
deadly gases killing us away.
Obliterating earth in a most unusual way.
Careless handling day after day.
Killing millions of animals along their way.
Why does it continue day after day?
Is it lack of knowledge?
Or just plain greed?

Source: Thinkstock Photos

We Did Not Notice

Animals were first to know.
Something was wrong;
something wasn't right.
Mother Nature was in trouble,
out of control.
For the first time in millions of years,
she was unable to correct her flight!
Lost her balance, our course.
Lack of ballast to keep us upright.
Powerless, with little control.
Our weather changing!
Developing larger, stronger storms,
seizing our skies.
The birth of super storms,
which are on the rise!
In the blink of an eye,
they sweep down from the sky.
Showing no mercy,
ready to fight!

Mother's Pump

Splash #1

Do you believe our ground is solid?

It's actually comprised of massive pieces of rock. Plates that float on top of a layer called the mantle. Where these plates unite are faults. The areas in between these faults are boundaries.

There are three types of boundaries: convergent, (creates pressure), divergent (pulls), and transform (pushes), creating a vacuum for the distribution of crude oil around the globe as needed for balance. I call this system "Mother's Pump."

Mother Nature distributes crude oil around the globe as ballast to keep us true to our axis, balanced upright. Mother's Pump works nonstop, 24-7, to offset our surface weight influences upon the earth, keeping us balanced.

Did you know there are over a million earthquakes per year around the globe? Plus millions more that are so small they are never reported.[1]

Continually she is pushing, pulling, pumping, and moving her crude oil, redistributing the weight of crude oil for balance.

Mother's Pump

Source: *US public domain USGA fault map*

Balance

Clash #1

Reflecting back when you were a child, remember that spinning top that gave you much joy? Spinning, spinning round and round. Yes, and then it fell down! Your parents would say, "It's all about the balance! Balance causes it to stay!"

Like your car tire, proper balance will keep things from going astray.

Did you know Earth is the same way?

Barrel after Barrel

Splash #2

A barrel of oil in the United States and Canada is defined as forty-two US gallons, 306 pounds.

The first oil well was drilled in Titusville, Pennsylvania on August 27, 1859.

Between 1859 and 1979, total oil production was approximately 135 billion tonnes.*

Between 1980 and 2013, we extracted approximately 308 billion tonnes globally.

To date, there have been five wellhead blowouts, causing 1.9 million tonnes to spill, killing millions of animals and plant life.

As of December 31, 2013, approximately 444.5 billion tonnes of recorded weight have been depleted from the earth ballast, causing Mother Nature (the earth) to shift on her axis. A large percentage of tonnes has been transformed and released into the atmosphere, polluting our air, depleting our ozone, and affecting the jet stream. The remaining oil has been stockpiled around the world, influencing our balance as we prepare for the day it's all gone.[2]

---

*Tonnes is the term used to describe the world standard of measuring oil.

Think about It

Clash #2

How can you state that extracting crude oil is not having an negative effect on our planet, unless you were born prior to 1859 (the birth of the first oil well) and can affirm that recordkeeping has been absolutely accurate?

Clash #3

So if an average car tire of forty-seven pounds can be perfectly balanced with a quarter-ounce weight, what effect do 433 billion tonnes of depleted weight have on our earth?

Clash #4

Do you know of anything on this planet to be just for us to exhaust without consequence? Without some type of reaction, positive or negative?

2
AWARENESS

Weather

Storms have been around for a long time.
A very long time!
Perhaps from the beginning of time.
Evaporation connecting earth and sky.
A water distribution system,
influenced by wind and clouds.
Not perfect but better than today.
Disparity within this system
has disrupted our climate's symmetry.
Too much water!
Not enough!
Climate balance consumed.
Global contamination has begun.
El Niño grows stronger.
Mother's pumping,
provoking El Niño to grow greater.
Oil replaced by the inner core,
Mother drawing more and more,
warming Pacific floor.
Evaporation more and more.

Riders in the Sky

Always trying to hitch a ride.
Teaming with other dwellers of the sky
increases their forces,
multiplies their size
whenever along for a ride.
Waiting to strike
when the pressure,
wind, and rain are right.
Without notice, day or night.
Sweeping down, wanting a fight!
Larger, stronger, they grow by the year.
Multiplying, expanding as they fly.
Hanging from other dwellers of the sky.
Transforming our cities, towns, our lives!
Blazing trails as they fly.
Whatever they touch, they crush,
showing no discrimination for property and life.
Taking what and whom they want.
Stay clear of the rider that hangs from high
when facing the dwellers of the sky!
Who knows what may appear.
Hurricanes, cyclones, typhoons,
tornados, tropical storms.

Riders in the sky

Source: Thinkstock Photos

So Much, Too Much

What to know? What to say
to someone who lost it all in a day?
What do you do?
What do you say?
Her husband, daughter, home,
everything she owns,
entirely swept away.
In a second, an instant, all is gone!
Not knowing,
wondering. Maybe they're okay?
What can I do? How can you say,
"It is going to be okay"?
It's all lost, gone, gone away!
Her life scattered every which way.
All her pain, her sorrow.
Words are not enough.
As she stands before me this day,
covered in blood,
her motionless son at her feet.
She drops to the ground, reaching for him.
Hopeless expression sealing her face!

Dear Lord, what do I say?

Cyclones, Hurricanes, Typhoons

Roamers, atmospheric consumers.
Petroleum gas feeders!
Searching for other dwellers,
riders of the sky.
Transforming into slow-moving,
colossal slayers roaming our skies.
Advancing from shore to shore,
influencing pressure, winds, and more.
Evolving like never before!
Do what you can. Do what they say.
Know what to do, know where to go.
Just get out of their way!

Roamers of the sky

Source: Thinkstock Photos

Develop a Family Emergency Plan

Splash #3

Knowing the weather risks in your area will help you better prepare for possible hazards. What would be your family's emergency plan?

Make the Time. Make a Plan.

Plan where your family would meet and how you would communicate. Consider that your family may not be all together when an emergency occurs. Take into account power outages, no cell service and no means of safe transportation. How would your family react to these different scenarios?

One such emergency plan might include having a set of ribbons; assigning a specific color to each family member. When a family member arrives at a designated rendezvous, if no other members are present, they would leave one of their ribbons marked with information about their whereabouts, so the family would be able to regroup at the next location.

This sample emergency plan is just a start to get you thinking. There are several emergency plans found on the web. Search for the one that best fits your location and your family's needs:

- local government emergency sites
- Google or YouTube
- www.emergencyreadyyourself.org

It is very important that every member of your family agrees to the plan and has a copy in his or her emergency kit.

Be ready. Have a plan.

Source: Thinkstock Photos

What's the Difference?

Splash #4

You may recall from grade school the four weather zones around the globe: the Atlantic, northeastern Pacific Ocean, western Pacific Ocean, and Indian Ocean.

The differences between a hurricane, typhoon, and cyclone have to do with their locations within the four weather zones. Different names identify them by regional location. In the Atlantic and northeastern Pacific Ocean, they are hurricanes. In the northwestern Pacific Ocean, they are known as typhoons. Cyclones occur in the South Pacific and Indian Ocean.[4]

To be classified as a typhoon, hurricane, or cyclone, a storm must reach wind speeds of 74 mph. When winds reach 150 mph or above, they becomes a super storm.

So why did Hurricane Sandy become a super storm with winds of 85 mph? What happened to the 150 mph rule for super storms? Was it Sandy's size, the amount of damage, the death toll, the barometric pressure, or the storm surge? Or maybe this monster storm is a new breed of storms?

All of the above!

The Hurricane Wind Scale was developed in 1970 by Herbert Saffir and Dr. Robert Simpson, the director of the National Hurricane Center. The current version is strictly a wind scale. Previous versions listed other factors, such as pressure, wind, and storm surge. It was modified in 2012 after Hurricane Sandy.

Today

Splash #5

Today, the Hurricane Wind Scale is a one to five categorization.

- Category 3 hurricanes now have a wind speed range of 111–129 mph.
- Category 4 wind range is 130–156 mph.
- Category 5 winds are 157 mph and above.

Although the intensity and frequency of active storms have increased globally 23 percent over the past six decades, we are also experiencing a global weather shift. The volume of storms in the Indian Ocean region is decreasing, while weather in the Atlantic Ocean region is expanding, setting record levels for precipitation, pressure, magnitude, and duration![5]

What's the Fuss?

What's the fuss?
This global fuss, fuss?
So much fuss.
Nothing new.
Why the fuss?
Just stay clear of the orange waste!
Don't enter the toxic zones.
Keep your mask on.
Don't drink the water.
Don't walk out on the open ground.
Stay out of the sun.
Our weather is changing.
Nothing new!
It does that from time to time.
What are you going to do?
Weather's gone bad,
doesn't bother us.
We're much too busy with all our stuff.
No need for such fuss.
We haven't begun to glow!

Wake up and look around!

3

PROFESSOR WHAT-IF'S STORY

Professor What-If's Story

It all started in September 2008. I awoke from a very restless night with an uncomfortable awareness that something was wrong—a helpless feeling deep within. This would continue to slowly build for the next two years until late April 2010.

I went through a period where I found myself up all hours of the night, tossing and turning, up and down, pacing the floor, back and forth in a state of nervous anxiety until the early-morning hours.

My danger alarm was going off; something was wrong, or someone was in trouble. I began to work my way through the family, processing my children, grandchildren, stepson, and close friends one by one. Checking Facebook for any clues that could reveal the identity of who was in need, I would look for something/anything that had maybe been hidden or overlooked. I found myself repeating this process, going back over older posts to find something, anything—a clue that would lead me to the identity or something that was about to happen.

I recall sitting up all hours of the night trying to recall recent conversations that would reveal a clue about what was going on. You know that nervousness that overcomes you, that feeling down deep inside that just possesses you? In all my years, I had never experienced this level of anxiety before. It was as if someone or something had me by the throat and was trying to tear my heart out! Why was this happening to me? What was I supposed to do?

After weeks of angst, I accepted my obsession as a form of guilt. You see, my first and only girlfriend, from age fourteen on, my wife of thirty-seven years, had passed away a few years earlier. I felt maybe she was trying to reach out to tell me something I should watch out for or something that was about to happen to one of the kids. Should I warn them? Or maybe it was me. Should I be on the lookout for something or someone? I felt like I was the one who had to be responsible. I was the only one left to look out for the kids. I needed to be on guard like she would have been.

During that five-month period between April and September, I experienced this unusual level of overwhelming anxiety that lasted for days at a time. My alarm would not turn off! Four of the five ended up to be false alarms, or so I thought at the time. Much later, I discovered what was going to change my life forever.

After each episode, I would wake up with a newfound knowledge of the earth. Yes, I know this sounds so crazy! Early on, I kept this to myself.

I don't know how or why this was happening or even how to explain it. I didn't hear voices or have any type of mystical happening. I would just wake up and know!

Then one night I was out for dinner at the yacht club with a few friends, and a special news report came on the television. Everyone was watching, listening intently. This is when I burst out into an explanation of what was happening and what was about to happen and why. Now that's not like me at all!

My friends were shocked; no one knew what to say! Everyone was staring at me, and then the room filled with a low whisper. I felt embarrassed. What was I saying? I wanted to crawl away, disappear into the night. Barbara, my wife, just stood there with her mouth open. How did I know this stuff? I am no geologist, and I had no formal background, so how did I know this stuff? But at the same time, I knew it was true! Extracting crude oil at the rate we're going, along with head-blowouts, oil spills, and leaking wells—it's destroying our planet and wildlife!

In the time that followed, I tried to keep myself under control, occasionally revealing measured amounts of information to close family and friends. Just enough to gain feedback and clarify this "ballast" theory in my own mind. I found that engaging in discussion seemed to broaden my knowledge. Somehow it allowed me to understand. I discovered this process helped me to create a bridge between fragments of this newfound understanding. Reflecting back, this seemed to be triggered whenever there was a report of an oil spill or a super storm somewhere.

The frequency of nervous, early morning hours seemed to pass. Oh, I still get those tossing and turning nights, worrying about my kids, but nowhere to the level of those nights back in April 2010 or the outburst I experienced that was triggered by the Gulf oil spill news report. Now I just wake up and know!

Thinking I was going crazy, I reached out for anything to help me understand this newfound knowledge. Was it real or not? I spent the next few years trying to prove it wrong.

Then, on February 14, 2014, a report came on the news about a major oil company preparing an agreement with Russia to start drilling for oil in the Arctic. I felt that high level of overwhelming nervousness rushing back. I knew I had to do something as soon as possible. But how, what … *I must write this book.* I was no author! *How do I go about writing a book without people thinking I am nuts or causing a panic?* So I started writing about my newfound knowledge and what I thought was wrong about extracting crude oil from the earth. I was one month into this process, with my head bulging with ideas, when one day I woke up and knew it was all wrong. I had to start over. The format was wrong; it read like a book of doom and gloom. My level of nervous anxiety continued to build, but I felt somehow guided. Four times, I restarted this book, researching way into the night. I went through this period when I would wake up from a sound sleep feeling like my head was about to burst! I had to offload some of this newfound information before I exploded!

My goal is to create awareness, with the hope that we will reduce or slow down crude oil production and stop supporting old petroleum-based technology. We must move funding from petroleum-based to renewable energy development, to control R&D pricing!

4

ENVIRONMENTAL IMPACT

Perplexity

Clash #5

While researching the environmental influences of the producers and contributors of global warming, I could not uncover two sources that appear to agree. Data seemed to be polluted, twisted, bent, pulled, and stretched in an effort to report fluff, with or without the facts, in an effort to look new and different. Somewhere along the way, we misplaced the reality of informational truth! What happened to responsible journalism and reporting the facts and pure news?

Yes, the truth has been infected!

Journalism has been replaced by sales and marketing greed ... infecting our print, television news, and even the web search engines!

We are now being led on a journey by marketing people, through a world of confusion and deception.

Watch how different networks report the same story. When you perform a search on the web, you're brought on a tour of landing pages that drive up revenue, leading you around and around in an effort to build up someone's landing page count.

Just watch the same news report by different networks, and you will see what I mean. No wonder we're so confused and can't agree on anything! We are trying to navigate our way through an enormous amount of conflicting information, in a sea of distorted facts!

My wife and I were bystanders at a shooting incident at the National Zoo in Washington, DC, and were locked down while the dramatic scene played out directly outside the building we took refuge in.

At the same time, we watched three major TV networks report this "breaking news."

I can tell you that while we watched these simultaneous reports by three different networks, only one was even close to accurately reporting the truth of the initial story, with about 75 percent accuracy by the third time it aired. The early reporting was a collection of misleading hype and garbage. No wonder we are so confused!

Nevertheless, it doesn't matter what process is contaminating our environment: manufacturing, transportation, or the generation of electricity. All combustion of fossil fuels is an immoral act! We're destroying our environment and everything in it. The sad truth remains; it is all driven by greed!

Deception

Clash #6

Could our confusion be a product of deception, or do we just choose to ignore the crisis?

Thus, I give you greed!

You know, I used to believe that 80 percent of the people would tell you 80 percent of what you wanted to know; the other 20 percent you needed to find out for yourself. But now I think it has reversed; 20 percent is truth, and 80 percent is fluff!

Confusion

What to believe?
Whom do you trust?
Confusion, deception,
misleading fluff!
Media's mission is to distract?
Intending to confuse, pollute,
injecting opposing angles, margins of truth
in this time of fluff and hype.
Stop the debates.
Put down your number.
Move beyond the confusion,
into the light.
Locate the truth!
Don't be distracted.
It's in between the lines.
Step across the greed into the truth!
Beyond the masters of fluff and hype,
where did the honesty go?

Green

A marketing label,
intended for food.
No chemicals, additives.
Preservative-free!
Nothing injected,
fresh off the vine.
Natural, organic, vitamins pure.
So why do we pay more?
What's the justification?
We know it's not R&D.
One hundred percent natural, pure.
Why does it cost more?
Marketing creation.
Awareness of green!
Or a marketing ploy?
Attraction of greed!

5

TRANSFORMATION

Flickers Day One

Faraway glimmers in the midnight sky.
Miniature flickers on a gray-black night.
Twinkle, twinkle, yellow glow,
affecting a few passers below.
Pausing, stopping,
standing in place!
Joined with the ground,
secured in faith!
Examining this twinkling yellow glow,
tears streaming down their faces.
"Out the way!"
fills the midnight sky,
echoes the passersby.
"Out of my way!"
broadcasts into the night!
Watchers locked in place,
passers rush into the maze.
"Transformation has begun!"
reverberates the midnight sky.
Watchers chant as one!
Suddenly embellishes the gray-black night,
sarcomas of a bright yellow glow.
"It has begun!"
echoes the watchers below.
Passers breaking free of the maze,
now in a run, scattering below.
Look to the left; a man in a long gray overcoat navigates the maze.
Seems to be taking notes!
Wandering about in a long gray overcoat.
From watcher to watcher,
what does he know?

What is this glimmering glow?

Flickers Day Two

No watchers standing below.
Normal movement.
Normal flow.
No glimmering lights.
No yellow glow.
Where did they go?
Up all night.
Watchers patrol.
Long gray overcoat!
Where did they go?
Does anyone know?
Must have dozed off.
Was it a dream?
Suddenly overcome with doubt.
Dark, angry cloud refocuses the light.
Difficult to see. Anyone there?
Violent winds begin to blow!
Off to my right in the fringe of the light,
long gray overcoat!
From last night!
Moving slowly, plotting his way.
Soon he's beneath my feet.
"Sir, dear sir … sir!"
I bellow below.
"Up here, above you!"
On the balcony overlooking the park.
Just then, a violent gust thrusts me back.
I awake on the deck.
Peering down through a space in the decking,
something gray flapping in the wind.

Flickers Day Three

The rain continues,
around noon I'll say.
Swollen skies, bursting rain,
like syrup flowing down.
Thick orange fog masks the park.
Watchers gather below.
One by one, they begin to show.
Clearly they must see the yellow glow.
Almost dark, they continue to appear.
Gathering in the park below.
Around the globe, reporting news.
"People standing in the streets,
watching a flickering yellow glow!"
There! Long gray overcoat with a black pocket on the right.
"Note taker!"
He's back. I know it's him!
Over to the right,
just off the street.
Must be his torn pocket flapping below.
He doesn't see me.
He's not taking notes, nearly running.
What's he clutching?
Running around the park.
He stops! Checking his notes.
Flash! A white light,
followed by a loud bang!
Bright white light, beyond the sky!
A green-yellow glow.
Clearing skies,
no rain, no wind.
Simply a bright, warm, flickering glow!

Flickers Day Four

Faraway flickers sprinkle the sky.
Mystical experience warms you inside.
A pallet of soft, melting colors
perform on stage in the sky.
Drawing us into some other place.
Warm, peaceful feeling fills you inside,
body, heart, soul, and mind.
People stand watching the show,
as if under some hypnotic spell.
Everyone happy, polite, and nice.
Peaceful, warmer, nice.
People singing, chanting, praying,
way into the evening glow.
A spray of miniature light fills the night.
Soft but bright.
Ungodly odor floods the clean night air.
Birds begin falling from the sky.
Another flash fills the sky.
Someone yells, "It's no spiritual door!"
Others join. "We're staring into hell!"
Floods the night sounds.
A doorway to what's to come.

Day Five

Breaking news: "A jet ignited petroleum gases
high in the sky, killing all on board on an
international flight."
Wildlife continues to fall.
Birds, insects, all!
No air traffic allowed.
No communication to be found.
All satellites have been destroyed.
Murky, orange glow fills the sky,
burning layers of gases we know.
Much higher than weather below.
Remaining gases continue to glow.
Thin layer of oxygen encases the earth.
How long, no one knows.
Raging weather circles the globe.
Toxic rain, winds, and orange snow.
Plains states evacuation underway.
Around the globe, stories flow.
Animal-like behavior rules.
Governments losing control.
Seas rising, wild fires burning.
Balance and order out of control!
Earthquake felt around the globe.

Earthquakes

Four weeks of quakes shattering our world.
Unsettled sensation suffering inside.
Vibration all around!
Shivering mind
rambles up and down.
Wobbly, wobbly ground.
Trembling inside shudders one's mind.
Everything falling down!
Climb to the top! Don't stop!
Sensations wearing me down.
Hard to walk, climb, climb to the top
unsullied position to survive?
On the roof, surely safe!
Objects plunging down,
swaying back and forth.
Nearly one with the building behind.
So much noise, trembling inside!
Dared to jump to the building behind.
Next pass!
Where did they go?
On the fringe of a hole.
Enormous cavity below!

Volcanoes

Volcano merges with earthquake.
Wobbly, wobbly ground.
Consuming buildings down!
Vanishing cities, entire towns.
Lava rains down.
Opening ground, dragging down.
Hollow oil cavity's down, down, down!
Gone, never found.
Beware of shaky ground!
Quivering seas, wobbly trees.
Submerged volcanoes swamp our seas.
Seafloor dwellers,
massive explosions.
Raging water curling down.
Molten rock propelling around,
now descending down.
Muddy ash covers the ground.
Mucky water raining down.
Hard to run on this wobbly ground.
Wall of water crashing down!

Maybe Our Future

Clash #7

I'm looking into a window, a magnetic fragment of time. For what is now, for what may become. A magnetic prediction into a window of time.

Soon we will have to surrender two of the Tornado Alley states as uninhabitable, and then our Great Plains states, followed by our coastal cities and towns. El Niño will boil our oceans! Our skies will become toxic, contaminating our food and surface water. Stay away from the rain and sun.

Beware of the winds and bitter cold!

Our seas will rise and fall to none.

Mother's Pump draws our center core closer to the surface. Our magnetic attraction increases, drawing meteors to rain down on earth.

Gases continue to burn in our atmosphere, consuming our oxygen. Gravity will increase, and then it will decrease!

Magnetic Storms

Oil is gone! Mother's Pump continues on!
Pumping, pumping, pumping,
pumping strong!
Stronger, stronger.
Pressure assembling along the coast,
pumping continues on.
Pressure building,
earthquakes intensify!
Triggering more volcanoes to turn on,
explosions around the globe.
Earthquakes intensify.
Sinkholes follow along.
Pumping goes on and on.
Pressure building,
she ruptures the core,
replenishing her oil with magnetic ore.
A magnetic reaction around the globe,
meteor showers raining down,
lured by the magnetic core.
North transferring all around,
dancing, dancing up and down.
No communication to be found.

Day by Day

All four seasons in a day!
Every day, much the same way.
Repeating the same conditions,
day into day, every day.
A muggy summer morning,
afternoon filled with fall.
Polar-like winter evening.
And that's not all!
Every few weeks,
night overcomes day,
no light for three to four days!
But now it's changing,
mingling each way.
Fall becomes morning,
wintery days,
sizzling summer nights,
with continuous light!
Multifunctional rotations, so they say!
How long can we go on this way?

Adjustment

Sitting in the darkness, gazing out my office window on the thirty-fourth floor.

My daughter Emily's twelfth birthday today. Someone shouts my name, "John, John," as a small light appears dancing around the room, followed by loud noises reverberating. A man steps in through my broken windows in a long gray coat. The note taker! "Where, where did you get that coat?" I ask.

"It's my old field coat. It has several large pockets … I used to wear it out on research projects," Bill responds.

"It's you! You're the man in the park! I have your pocket! Why are you here?" I ask.

"The same as you, John. I work in the lab on the fifty-eighth floor. How do you know my name?"

"We met briefly in the park several years back, and you bought me a coffee after I spilled the one I had all over the elevator that day. You're Bill Wo; you're the note taker! Well, I didn't know you were the gray coat man!"

We were not allowed to discuss our projects.

"What were you doing that day in the park, Bill?"

"I was working on the Sky Project," Bill responds. "But right now we need to focus on getting back to the surface. I'm afraid we fell fairly deep!"

I ask, "What happened? How did we get here?"

"Not sure, but by the smell of it, I think the entire building fell down into an old oil cavity. We were sucked down during that last earthquake. But we need to go now. I need your help setting up the carbon collectors and oxygenators in my office to replenish our oxygen. Sorry, we'll have to climb up the outside of the building access; within the building is blocked."

Surrender

Submission has begun!
Sky, earth, and sun.
Swirling sands, raging seas.
Rain on overflow!
Hottest, coldest, hostile times.
Elements capture control!
Global balance shattered.
Mother loses control.
Once our benefactor,
now our nemesis!
Riders seize control,
destroying our host.
As we debate, why?
Climate change takes control.
We couldn't agree,
overcome by greed.
Now we have to leave.
No longer welcome,
exhausted our stay.
Not safe for what's to come.
Run, run, run.
It's time to go!
Now!

What Went Wrong

Clash #8

When the crude oil levels drop below Mother Nature's ability to distribute, Mother's Pump continues to pump, trying to distribute crude oil to correct our balance weight. This action triggers numerous earthquakes and volcanoes. Old oil cavities collapse into sinkholes around the globe, consuming cities and towns. Volcanoes rise up from under the sea, spilling poison gas. Pressure continues to build, fracturing our inner core, replacing crude oil with magnetic malted ore. Amplifying our magnetic attraction to objects in space. Giving birth to new types of magnetic storms. Causing us to move underground!

6
LIVING UNDERGROUND

Say Good-Bye

Welcome to global climate domination.
Atmospheric paroxysm.
Intensifying beyond belief.
Say good-bye to life on top.
What once was is gone!
Sinkholes swallowing cities,
homes burning down.
Gases coat the air,
debris propelling around.
Mother has lost control;
global damnation reigns on.
We messed it up!
Discussed too long,
couldn't agree.
Blaming everything but us.
There's no turning back.
All life is gone!
We're the last to go.
No longer safe.
We no longer belong!

On Top

Gram-Papa!
Tell us about living on top.
Kids playing in the park
until way after dark!
To run in the wind
with the sun in your face.
The smells of spring flowers
after a warm rain.
The sound of a bird's peep?
The wind in the trees,
rain in your face.
Swimming in a lake!
Children playing, laughing.
Tell us about strolling after dark.
A star-filled sky.
What about the porch light?
Riding a bike.
The breath of fresh air!
Gram-Papa, Gram-Papa,
what's it like?

Mom

Dad said you used to lie in the sun!
Play around and have fun!
Day into night, what was it like?
That was before.
Before what?
Before you were born.
Before the super storms!
The period of the killing winds.
Just after the orange rain.
When everything turned brown.
The freezing cold.
No food, poisonous water.
Before we moved underground,
the greedy ruled, created this dome!
So many people lost their lives,
victims of the killing storms!
Burning sun, raging water, sinkholes.
No one would listen
when warned of climate change.
Back when everything was a debate!
When greed ruled.
By the time they agreed,
it was too late!

Reflection

Do you recall those warm summer nights,
walking along the bay?
The humidity seemed to just fade away!
Our strolls on the beach
after a hot, steamy day?
Cool summer breeze.
Moon-guided tours.
The melody the waves would play?
I recall the moonlit shadows,
reflecting on your face.
Highlighting your body
in a most shimmering way!
Rolling in the sand.
Making love in the bay
made everything fade away.
Perfect summer nights.
Lying nude next to you
until the break of day.
Remember those games we would play?
Did you ever think it could all go away?

Once Was

You open your window to the cooling breeze after a hot summer day. Gazing at a star-filled sky, inviting you to step outside. Strolling out the door along your porch, pausing to enjoy the cooling breeze. Along the garden, the fresh-cut lawn, moonlight guides the way to your favorite place that dissolves a bad day. Looking out over the bay, reflecting stars dancing away, the sky so happy today. A gift from the heaven above.

Mom, Mom, wake up!
Time to dig that other tunnel!

Time

Time to encourage,
influence change.
No more arguments.
No more blame!
End the discussions,
disputes, and blame!
Not a contest!
Not a game.
Stop the debates
before it's too late!
Time to move on,
time for change.
Stop the mist of confusion.
Let go of your greed.
Stop government subsidies.
Say no to petroleum greed.
Think of the future.
Time of discovery,
exploration, change!
Time to develop
a new energy way.
Time for change!

What If

It's more than we can conceive.
More than we will believe?
What if?
Petroleum was the shield
of the magnetic core?
The gravity door!
Controls the solar door!

7

PROFESSOR'S INSIGHTS

Flash

What It All Means

Section 1

The Book of Facts, Poems, Riddles and Rhymes: Represents the format of your journey.

Need for Greed: Represents the consumption and need for gasoline and petroleum-based products.

Mother Nature: A cry for help. Lack of petroleum has affected our solar balance, causing our planet to wobble. This waffling is influencing our polar vortex, causing a vacuum and drawing El Niño closer to the surface, warming our oceans and creating a major evaporation zone to reoxygenate our skies.

Losing Ballast: Crude oil is the earth's ballast, necessary for balance.

We Did Not Note: Ignoring the signs of change. How we continue to support old petroleum-based technology. Our failure to develop new technologies.

Mother's Pump: The distribution system of oil around the globe. How the plates and boundary are used to reposition the weight of crude oil continually around the globe, creating ballast. Insufficient crude oil levels have fractured the inner core, causing Mother's Pump to draw molted ore within a cavity under the sea, called El Niño.

Balance: A proposition that it only requires a very small percentage of weight to affect balance.

Barrel after Barrel: Defines the volume measurement of a barrel of crude oil and the quantity of crude oil extracted from the earth over the last 153 years.

Think about It: Proposes the question of impact.

Section 2

Weather: Deadly gases invade our sky. Inefficient crude oil pressurizes Mother's Pump to rupture molten lava flow. Warming Pacific Ocean floor adds evaporation zone. Balance is out of control. Too much water or not enough disrupts our atmosphere as we debate why.

Riders of the Sky: Tornados using other types of storms to move them around, drawing energy and power from the host storm.

So Much, Too Much: About a thirty-two-year-old mother and wife who lost it all within twenty seconds, just barely clinging to life. No money, no family, no home, and nowhere to go.

Cyclones, Hurricanes, Typhoons: Describes dwellers, roamers of the sky, searching for riders in the sky. Develop a family plan!

Develop a Family Emergency Plan: Offers information and consideration to discuss with your family and where to go for help.

What's the Difference: Information regarding our global weather zones.

Today: Current hurricane wind scale. Storm history around the globe.

What's the Fuss: Portrays a skeptical person without a care. Much too busy to notices change. A wake-up response from the professor!

Section 3

Professor What-If's Story: Explains the inspiration behind this book.

Section 4

Perplexity: Bewilderment as data has become polluted over time. Influenced by change and corrupt people. A real-life example of media hype.

Deception: Professor's comment regarding confusion and deception.

Confusion: Media's effort to inject misleading facts to confuse and invoke controversy.

Green: An example of manipulative greed. Misleading deception of our food source. What will they say going forward regarding the effects of climate change and how it contaminates our food source?

Section 5

Flickers Day One: Fear of the unknown. The power of people, curiosity, spiritual need, and impudence.

Flickers Day Two: Lack of trust in oneself. Fear of the unknown.

Flickers Day Three: The power of numbers and the need for knowledge.

Flickers Day Four: Mystical guidance, spiritual hunger, truth revealed, reality takes control.

Day Five: Realization we have gone too far. A need for understanding. A fear of what is to come!

Earthquakes: Stages of fear, questions, survival, and disbelief.

Volcanoes: Survival, uncertainty, hopelessness.

Maybe Our Future: Professor's vision of the future. The stages of climate change.

Magnetic Storms: Mother's Pump is causing the increase of earthquakes and volcanoes around the globe. Replacing crude oil with magnetic ore for ballast, affecting our magnetic influence on and around the globe. This interference is also affecting our global weather.

Day by Day: The disruption of balance, multifunctional rotation. The effects of climate change.

Adjustment: One morning, John arrives at work early, hears a loud noise, and wakes up miles underground on his daughter's birthday. He discovers the note taker with the long gray coat is Bill Wo, a research scientist who works in his building.

Surrender: The elements take control, and climate change rules. Unsafe for humans.

What Went Wrong: Professor shares his foresight of cause and effect.

Section 6

Say Good-Bye: The remaining public surrender and agree to move underground, as it is no longer safe on top.

On Top: Years later, their children who have never set foot on top start to question the elders about what it was like.

Mom: Children question what it was like to lie around and not work. Trying to understand why they moved underground.

Reflection: A husband and wife recalling the past and what it was like on top.

Once Was: A woman dreams of the past as her son awakens her.

Time: Professor wants to encourage change. Explore new energy sources. Time to change!

What If: A glance into the distant future!

I know we can't stop our dependency on oil overnight. It has gotten beyond our control. Accept that we are all part of this shared responsibility, and now it is time to control our greed. It is time to work together and restore our planet.

Climate change is Mother Nature trying to repair herself, and we are caught in the process!

Share your comments with the professor at www.professorwhatif.org.
Join the discussion on Twitter: @profwhatif.
Contact the professor at: profwhatif@gmail.com.

Professor What-If
Self-Proclaimed

Notes

1. Source: Earthquakes
    <u>U.S. Department of the Interior</u> | <u>U.S. Geological Survey</u>. USGS-authored or produced data and information are considered to be in the US public domain.
2. Oil Production Sources: Barrel after Barrel / Splash #2. The World Crude Oil Production (by year) @ http://www.indexmundi.com/energy.aspx? product=oil&graph=production.
3. Oil Production Sources: Barrel after Barrel / Splash #2. US Energy Information Administration/International Energy Statistics: http://www. eia.gov/
    http://www.eia.gov/cfapps/ipdbproject/IEDIndex3.cfm?tid=94&pid=57&aid=32.
4. What's the Difference? Splash #4. NOAA's National Weather Service.
5. Source: Today / Splash #5. NOAA's National Weather Service Storm Prediction Center Annual Summaries.

The professor shares the experience of his existential awakening and it is one that might also be described as an ecological spiritual awakening.

His "awakening" is an enlightenment we should all have or strive to experience. We are beyond the debate. Climate change is real.

—Pacific Book Review

Printed in the United States
By Bookmasters